THE WORLD OF HOBBIES

HOBBIES IF YOU LIKE COMPUTERS

by Christopher Forest

BrightPoint Press

San Diego, CA

an imprint of ReferencePoint Press, Inc.
Printed in the United States

For more information, contact:
BrightPoint Press
PO Box 27779
San Diego, CA 92198
www.BrightPointPress.com

LIBRARY OF CONGRESS CATALOGING-IN-PUBLICATION DATA

Names: Forest, Christopher, author.
Title: Hobbies if you like computers / by Christopher Forest.
Description: San Diego, CA: BrightPoint Press, [2025] | Series: The world of hobbies | Includes bibliographical references and index. | Audience: Grades 7-9
Identifiers: LCCN 2024013360 (print) | LCCN 2024013361 (eBook) | ISBN 9781678208806 (hardcover) | ISBN 9781678208813 (eBook)
Subjects: LCSH: Computer science--Juvenile literature. | Computer programming--Juvenile literature. | Social media--Juvenile literature. | Internet--Juvenile literature.
Classification: LCC QA76.23.F6675 2025 (print) | LCC QA76.23 (eBook) | DDC 004--dc23/eng/20240402
LC record available at https://lccn.loc.gov/2024013360
LC eBook record available at https://lccn.loc.gov/2024013361

CONTENTS

AT A GLANCE

- Coding involves using written languages to program devices. A person may code a computer or digital device to complete a task.
- Beginners can learn how to code by watching online tutorials or taking coding classes.
- Social media gives people a place to communicate and share information with others. Many people use social media sites such as Instagram or TikTok.
- Social media users create content such as photos or videos. They may need to set up a posting schedule and engage with their online followers.
- People can film and edit their own videos. They post them to video sharing sites such as TikTok and YouTube. Some people use special equipment, such as cameras and video editing software.

- Gaming involves playing video games for fun. Some gamers are part of professional esports teams that compete in gaming tournaments.

- Gamers need equipment such as a computer to play on, a gaming console, a headset, and a keyboard.

- Many people enjoy creating websites online. People can use website building tools such as WordPress to design their own sites.

- Many websites include headers, a menu, and web pages. Website creators may write and brainstorm engaging content.

NOT JUST FOR HOMEWORK

Ben sat at his computer looking at the screen. His little sister, Li, walked into the room. "What are you working on?" Li asked her brother. "Just doing some work," Ben said.

Li looked at Ben's computer screen. It was filled with a bunch of symbols. Some were blocks with shapes. There was also a flag, an image of a rocket, and instructions. "Is that homework?" Li asked.

People who learn coding can make programs such as computer games and apps. Some use coding platforms designed for beginners.

"Sort of," Ben said. "We are practicing coding for a technology class at school. I am using this online site to make a **program**. It's going to be a computer game."

"That looks like a lot of work," Li said. Ben laughed. "It is," he said. He typed a few more instructions on his computer. "But it is actually a lot of fun. Let me show you."

MAKING A GAME

Ben pressed a key on his keyboard. The computer screen went blank for a moment. Then the game program he had made began to run. A rocket ship appeared on the screen. The player had to guide the rocket through an asteroid field.

Ben told Li about how the game program worked. "I move the ship using the

Some people practice their coding skills by programming small robots.

keyboard," he explained. He pressed the "up" and "down" keys. This made the rocket move up and down. The "A" key made the rocket move faster. The "Z" key slowed it down. Ben landed the rocket safely on a planet. He didn't hit a single asteroid.

Li was impressed. "You made this program yourself?" she asked. Ben nodded. "There is a lot you can do with computers," he said. "They are a great tool. Let me show you how to make your own game."

CHAPTER ONE

CODING

Coding involves writing code, or a set of instructions that tell a computer to do a certain task. These instructions are known as a computer program. When a computer turns on, for example, it displays a home screen. A computer program makes the computer show that screen. Programmers write the code to make that work. The person using the computer does not see the code. Only the programmers do.

Learning coding languages helps people better understand how computers work.

www.

Some people code for fun. They might make video games or program small robots. Others code as a job. They learn to write special coding languages. This code tells a computer or **application** what to do.

There are many types of coding languages. Binary languages, for example, use numbers. Other coding languages

Most coding languages are made up of letters, numbers, words, and symbols.

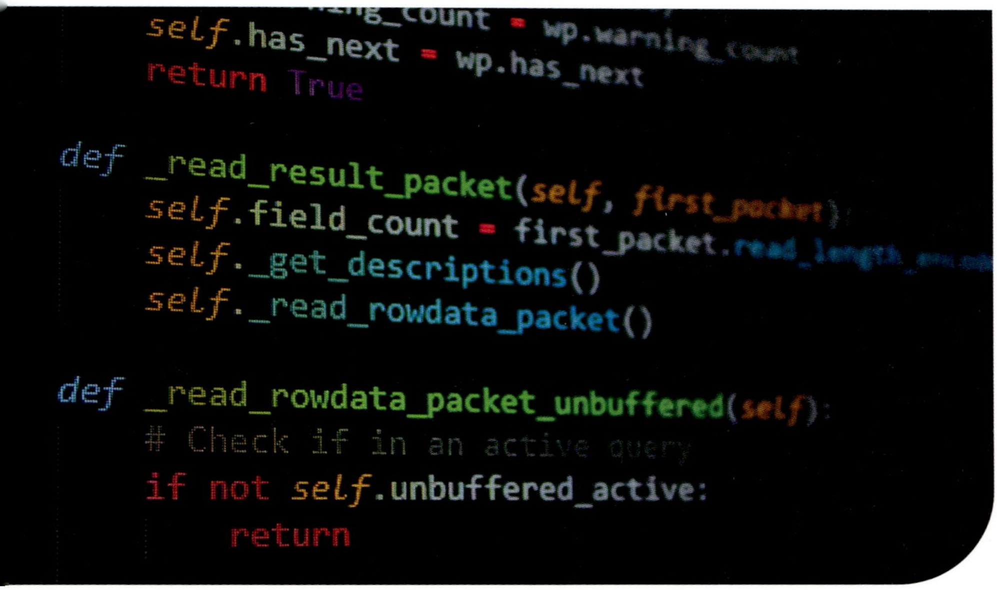

are complex. Examples include JavaScript, HTML, and Python. People use these languages to create games, websites, and more. Scratch is a simple programming tool. It helps beginners learn to code without coding languages.

TOOLS AND SKILLS

Beginners need a few tools to start coding. First, they need a computer that can be programmed. A keyboard and mouse may also be needed. Beginners must also learn a coding language. Websites such as Code.org or Scratch help people learn the basics of coding.

Coding requires several skills, too. One is creativity. A programmer must know what they want a device to do. Then they can write code to make the device complete that task.

Programmers must also be good problem solvers. Sometimes a device does not respond to code the way a programmer wants it to. When this happens, the programmer must **troubleshoot** the problem. Patience is important. A computer program doesn't always work the first time. Programmers may need to rewrite code in a different way to get the program to work correctly.

LEARNING TO CODE

There are many places where people can learn coding. Websites offer courses for beginners and advanced learners. Many schools offer coding classes for students, too. Students practice using code to make games or simple web pages.

Students often begin by using block coding. These are blocks of text that can be moved around on a computer screen. The blocks tell the computer what to show on a screen or what to do. After learning this, students can try more difficult codes.

There are many coding camps or groups that people can join. Colleges and

Hour of Code

The Hour of Code is a worldwide event sponsored by Code.org. People who participate spend one hour learning how to code. Millions of students across 180 countries have participated in the Hour of Code. The event is held during Computer Science Education Week every December. But students can also participate throughout the year.

Coding websites such as Scratch can be used for many projects. Some people use Scratch to program small robots or drones.

community centers usually offer them. Joining a group can be a good way to learn with other people who enjoy coding. Coding with others can be helpful. People can troubleshoot together. They can work together to make programs and games.

One well-known coding group is Girls Who Code. It encourages girls to learn coding. Student Callie Athon is part of a Girls Who Code club. She says, "It's interesting to figure out what people do to make all the technology work that I use every day."[1]

BENEFITS OF CODING

Coding has many benefits. It can improve a person's problem-solving skills. It can help them understand technology better, too. Learning coding skills can also be helpful in school. Bill Gates is the founder of the company Microsoft. He thinks coding strengthens people's minds. He said, "Learning to write programs stretches your mind, and helps you think better."[2]

Many high schools offer computer science classes. Students learn coding and programming basics.

Some coders turn their hobby into a rewarding career. In fact, the number of coding jobs is expected to grow in the future. Learning to code can lead to job opportunities. Coders may find jobs as web developers or computer programmers.

CHAPTER TWO

SOCIAL MEDIA

Social media is a popular hobby for many people. It involves communicating with others and sharing information in a digital space. Social media **platforms** allow people to connect with each other online. They send messages or post pictures. They share stories with friends or family. Many people even use social media sites to create an online presence. Some social media users post

People who enjoy posting on social media often take photos on their smartphones. Selfies are one popular type of photo shared on social media.

TEEM

Some Instagram users focus on collecting likes on the photos they post. They may add hashtags or clever captions to the photos to attract followers.

about their lives. Others post about specific topics, such as fashion, travel, or sports.

GETTING STARTED

To get started with social media, people must have a few basic tools. First, they need a device on which they can access a social media platform. This might be a laptop, tablet, or cell phone. Second, a person must join a platform. They may need to sign up for an account. Beginners can choose from many platforms. Some are designed for children under age thirteen. Popjam, Grom Social, and Go Bubble are three examples. Other social media sites, such as Instagram and TikTok, are meant for a wider range of audiences.

Beginners must also decide what type of content to make. This can affect what type of gear they need. For example, Instagram focuses on sharing photos. A person who wants to post high-quality photos can use their laptop or cell phone camera. As a person learns more, they may want to use a professional camera. They may also need a camera tripod. TikTok focuses on sharing short video clips. To make this kind of content, a person must learn how to film and edit videos. People can learn these skills through classes or online tutorials.

PLANNING AHEAD

Once a social media creator has their content ready, they can set up a

When filming videos, TikTok users may use camera tripods or ring lights. Some participate in viral trends such as dance challenges.

posting schedule. They might decide to post every Friday, for instance. Having a schedule can help social media users gain more followers. It can also help them build a strong online presence.

Style blogger Chloe Alysse uses Instagram. She recommends planning ahead and breaking things down into small tasks. "On day one, I'll take my outfit photos for the week and maybe capture a few extras to stockpile for a rainy day," Alysse said. "On day two, I'll work on captions."[3]

BENEFITS OF SOCIAL MEDIA

Using social media can be more than fun. It can help people develop useful skills. Amy Jo Martin is a podcaster and social media expert. She thinks social media "gives a voice and a platform to anyone willing to engage."[4] Making social media content helps people learn how to communicate well with others. They learn how to send messages and share ideas.

Some people also think social media can make people more creative. Many people post fun dances, engaging stories, or silly videos. They enjoy coming up with new

Staying Safe

It is important to stay safe while using social media. People should never post personal information, such as a bank account number. Someone could steal this information. Social media users should avoid sharing their location, too. Social media can also have a negative effect on mental health. Many people set a timer to limit how long they spend on social media apps.

Some people believe using social media can help people build relationships and improve their communication skills.

content ideas. People who use social media are often skilled at writing, too. They must make sure information in their posts is clear and easy for readers to understand.

Some people become social media influencers. These are people who have

Social media influencers may focus their content on specific topics. Some popular content categories include food, travel, and fashion.

built a large online following. Others pursue careers in social media. A person might work as a social media manager for a business. This involves overseeing the business's social media accounts. They create content to post on various platforms.

CHAPTER THREE

MAKING VIDEOS

A good hobby for people who like social media is making videos. The rise of social media has changed the way people communicate. Platforms such as YouTube and TikTok have made sharing videos popular.

People film videos with cameras or their smartphones. Then they edit the videos before posting them online. But not

People who make videos may use equipment such as microphones and ring lights. These tools help make videos look and sound professional.

everyone makes videos for social media. Some people make their own movies.

For many people, making videos is more than just a hobby. They have become successful YouTubers or TikTok stars. These stars are influencers. They post high-quality videos on all kinds of topics, from dancing to product reviews.

The Rise of YouTube

YouTube was launched in February 2005. It was developed by Steven Chen, Chad Hurley, and Jawed Karim. They wanted a reliable way to host videos online. They got tired of videos not playing correctly in texts or emails. So they created YouTube. Today, people all over the world post videos on the site.

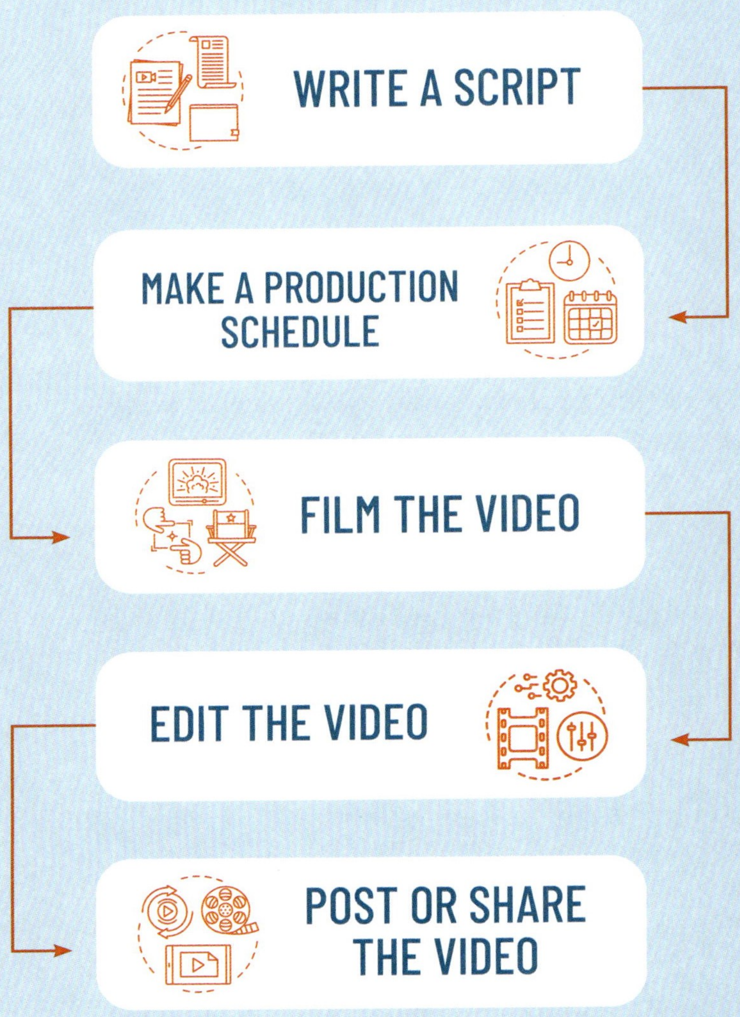

The process of making videos involves several steps. Many people start by coming up with an idea and writing a script. This helps creators get ready for filming.

TOOLS OF THE TRADE

Anyone can make videos for fun. It requires only time and a few tools. Daniel Middleton, known as DanTDM, is a YouTuber and video game commentator. He says, "The best

Video editing software allows users to cut and rearrange different video clips. Users can also adjust the video's sound and add special effects.

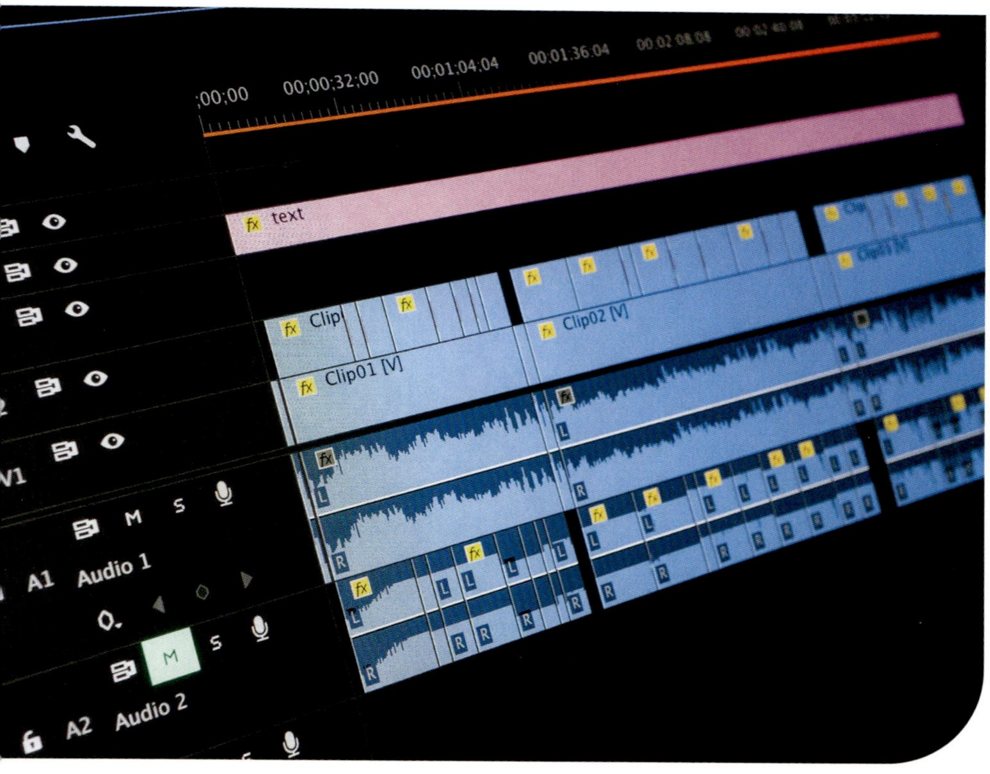

thing about YouTube is that anyone can do it, and that's exactly what I did."[5]

To get started, video creators need some specific equipment. First, they need a camera to record videos. This may be a professional filming camera. But people can also use their laptop or cell phone camera. They often need to write a **script**.

Video creators also need a computer with video editing software. This type of program allows a person to upload and edit videos. They can edit their video's sound and length. They can delete mistakes or add music. Some people add special effects. Many programs are free to use. Some options include iMovie and DaVinci Resolve. After editing their video, a creator must choose where to post it. YouTube and

TikTok are two common video hosting sites. Users can set up free accounts on these platforms.

LEARNING OPPORTUNITIES

Learning to make videos takes time and practice. The more videos a person makes, the better they become at it. Beginners can learn by watching tutorials on online video hosting sites. Many schools offer video production classes for students, too. People can even attend summer camps to learn how to make short films. By taking a class, video creators can work with others who have the same hobby. They can share video content ideas and skills.

Student Aiyanna Randolph attends a high school focused on film and television.

Many high schools and colleges offer video production classes. Students learn how to use film equipment and editing tools.

The school helps students build the skills they need for filmmaking careers. "I really enjoy writing stories, creating scenarios and stories for us to film," said Randolph.[6]

BUILDING VIDEO SKILLS

Making videos involves several skills. Video creators must learn how to set up and use

Making videos is one way for people to share their passions and interests with others.

video equipment. They must also learn to organize the time needed to shoot a video. They become skilled at outlining stories they want to tell. Video creators also develop strong editing and technology skills.

Making videos for fun can lead to future careers. Some people become social media influencers. Others create their own home studios to make independent movies. Video creators may find careers in television production or **screenwriting**. They may operate cameras or work on a film crew.

CHAPTER FOUR

GAMING

Gaming involves playing video games. People can play games on computers, tablets, **consoles**, or cell phones. They can choose from a wide variety of games, from open-world fantasy adventures to battle royales. Gamers may play alone or compete against other players. Some games are short. Others can be played over a long period of time, with new journeys happening as part of a game.

Some people live stream themselves playing video games. They use a camera and a headset to capture their reactions to the game.

Many people are casual gamers. They play video games for fun. Others are professional gamers. They play games as part of tournaments or contests. Many **live stream** themselves playing games on social media sites.

GETTING STARTED

To get started with gaming, gamers need some basic equipment. First, they must

The World of Esports

Esports is short for "electronic sports." It refers to video games played as part of a competition. Gamers compete as individuals or as part of a team. They often play in large tournaments and sometimes win prize money. Some esports tournament games include *Madden NFL*, *Fortnite*, and *Overwatch*.

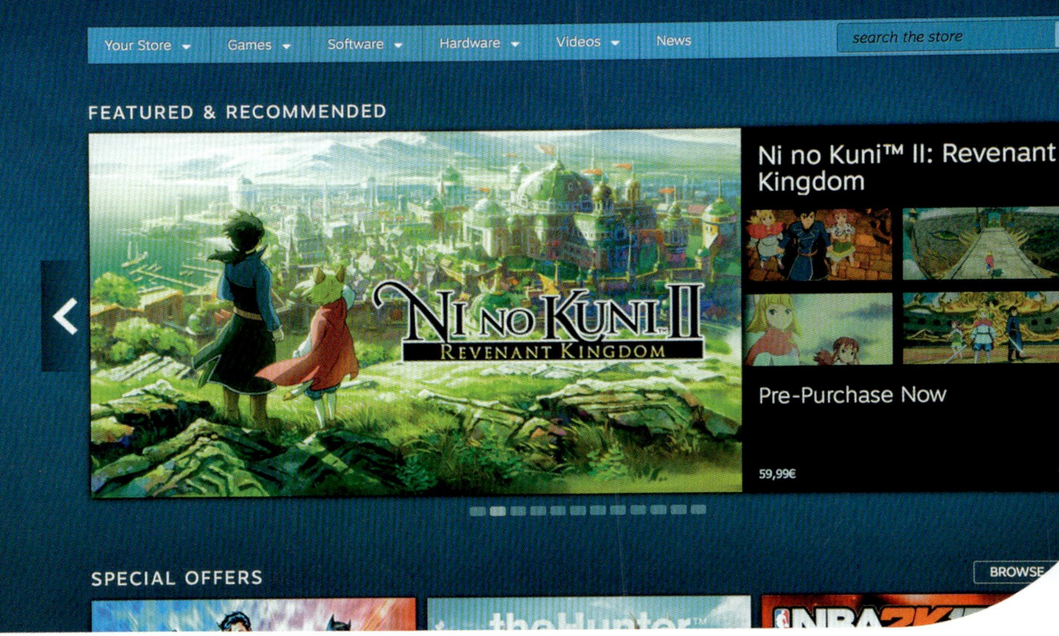

Gamers can access thousands of games on the Steam platform. They can buy games, create a game library, and chat with other players.

have a device to play on. Some games can be downloaded or accessed on a computer. Gamers can visit online video game stores such as Steam or itch.io. To play some types of games, gamers must purchase special consoles. Examples include PlayStation or Xbox. Gamers may also need a keyboard, game controllers, and a headset.

Gaming can help people build problem-solving skills. Some games involve solving puzzles or coming up with strategies.

GAMING SKILLS

People can learn how to game by watching videos or live streams of others playing online. They can even join gaming groups. But the best way to become better at a video game is to play it many times.

The more a person plays a game, the more skilled they become at playing it.

Gaming requires several skills, too. One is hand-eye coordination. This is the way a person's hands and eyes work together to complete a task. Gamers use this skill to quickly maneuver game controllers.

Gamers also need problem-solving skills. They often need to react to obstacles in a game. Sometimes they need to solve puzzles to progress to the next level. Problem-solving helps a player overcome these obstacles.

BENEFITS OF GAMING

Gaming can have several benefits. One is that it can help people feel connected with others. Lisa Su is CEO of Advanced

Gaming can be a good way to spend time with friends and build new relationships.

Micro Devices, a computer technology company. She says, "Gaming brings people together."[7] People often play games such as *Fortnite* or *Minecraft* with their friends. This makes gaming a very social hobby. Even when gamers are not in the same room with each other, they feel connected. They are part of a community.

Many gamers also learn teamwork skills. In some video games, they must cooperate with other players to achieve a common goal. Players must know how to work as a team. Some gamers form teams and enter competitions together. Irish gamer Emzii participates in tournaments with a team. She says, "Competing and the rush of winning is the best thing. But then also meeting everyone else. You are surrounded by like-minded people. We have grown to be like a family and a little community."[8]

Some gamers turn their hobby into a career. They become professional gamers. Others are interested in designing video games. They can find jobs as game developers. This involves coding, testing, and writing games.

CHAPTER FIVE

CREATING WEBSITES

People who are interested in the internet may enjoy making websites. A website is a collection of web pages located on the internet. It is often attached to a single location, or domain. A website can be accessed by typing its Uniform Resource Locator (URL) into an internet search bar.

Today, people create all kinds of websites. They may design a website focused on a specific topic. Some people

People can create all kinds of websites. They may work with others to design and maintain a website.

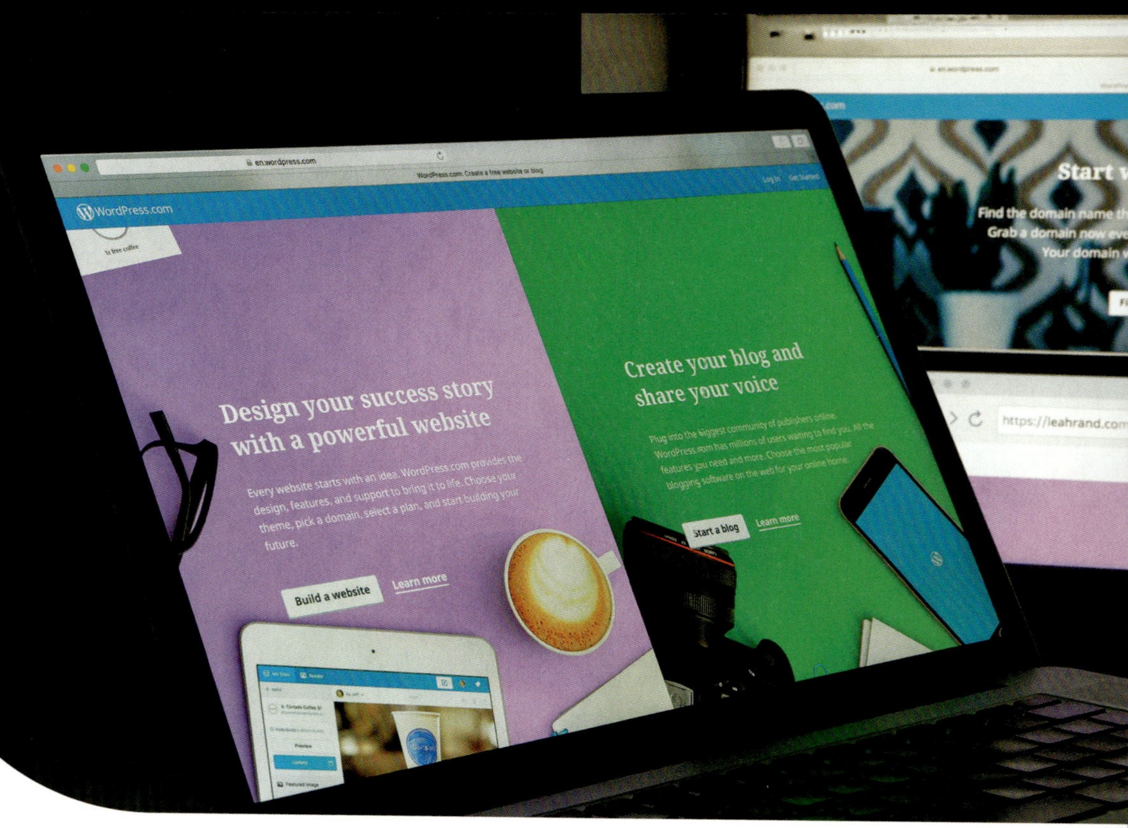

Sites such as WordPress offer easy-to-use design tools. Beginners can use premade templates and themes to build their sites.

are bloggers. They post updates about their lives or other topics on their websites. Others use websites to showcase their work. Authors and artists often have this kind of site. Businesses use websites to sell items to people. The possibilities for websites are endless.

WEBSITE BASICS

Some people use special coding languages such as HTML or Java to make websites. But writing code is not required to make a website. Today, beginners can find many website-building tools available online. Sites such as WordPress make building a website easy. These sites provide templates to help beginners design web pages without having to code.

HTML

HTML stands for HyperText Markup Language. Programmers use this language to create websites. HTML usually includes basic commands that programmers learn. These commands tell a website what to do. They also tell it what to show on a web page.

To get started, beginners need a few tools. First, they must have a computer or tablet. Then they must join a web hosting site. This is how their website gets a URL address and a domain name. This allows people to visit the site. WordPress and Google are two common web hosting sites.

Next, a creator must set up the site. Most website designs include a few basic parts. The top of the website usually has a header or title. This is often the first thing people see when they visit the site. Because a website usually includes multiple pages, a menu helps visitors navigate the site. This is a list of other pages people can visit on the site. They click on items in the menu to go to those pages.

For example, many websites include an "About" page that explains the website's purpose. Websites may also include a footer at the bottom of the page. The footer might contain links to the creator's social

Some people draw maps or plans of all the pages on their website. This helps them better visualize the layout of the site.

media profiles. The creator may also write blog posts to share on the site.

BENEFITS OF CREATING WEBSITES

Becoming a website creator helps people develop useful skills. Website creators must be creative and have a good eye for design. They must be organized, too. This helps them make a site that's easy for people to navigate. Many creators also post content regularly. They create a posting schedule to help. They brainstorm engaging content ideas.

Fashion blogger Courtney Quinn runs the website Color Me Courtney. She wrote an article about blogging for *Teen Vogue*.

Some website creators develop graphic design skills. They use graphic design software to design their own logos and graphics.

"You don't have to be what the industry considers an 'influencer' to make an impact on someone's life," Quinn said. "Just put your own unique twist on it—in the real world with kind acts, or online with beautiful images or words."[9]

CLASSES AND CAREERS

Website enthusiasts can build their skills by taking web development classes. These are offered online and at schools. Some local community centers offer them, too.

Beginners can take web design classes at schools or colleges. They learn how websites work and how to create them.

Creating websites can also lead to future careers. Many companies hire web developers to create and maintain their websites. Website creators may also find jobs in digital marketing. This involves advertising products or services online.

GLOSSARY

application

a computer program or software designed to help a person with a task

consoles

electronic hardware devices that run video games

live stream

to broadcast a live video on the internet for people to watch

platforms

websites or apps on which users can share, post, and read content

program

a set of coded instructions that tell a computer what to do

screenwriting

the practice of writing scripts for movies or TV shows

script

the written text of a TV show, short film, or movie

troubleshoot

to experiment with different ways to solve a problem

SOURCE NOTES

CHAPTER ONE: CODING

1. Quoted in Amber Brophy, "Girls Who Code Club at Woodland Inspires Students," *Johnson City Press*, May 1, 2023. www.johnsoncitypress.com.

2. Quoted in "Leaders and Trendsetters All Agree on One Thing," *Code.org*, n.d. www.code.org.

CHAPTER TWO: SOCIAL MEDIA

3. Quoted in Kristen Dahlin, "Confessions of an Instagram Influencer," *Tailwind*, October 14, 2019. www.tailwindapp.com/blog.

4. Quoted in "Social Media and the Illusion of Familiarity." *Medium*, June 4, 2022. https://medium.com.

CHAPTER THREE: MAKING VIDEOS

5. Quoted in Laura T. Coffey, "Millions of Kids Adore This YouTube Star," *Today*, October 12, 2017. www.today.com.

6. Quoted in Kate Sequeira, "LAUSD's New Magnet Film School Gives Students Hands-on Learning about Entertainment Industry," *EdSource*, February 8, 2023. https://edsource.org.

CHAPTER FOUR: GAMING

7. Quoted in Dean Takahashi, "AMD CEO Lisa Su Unveils 7-Nanometer Radeon VII GPU," *VentureBeat*, January 9, 2019. https://venturebeat.com.

8. Quoted in "Emma—The Inspirational Gamer Breaking Barriers," *GenderGP*, November 20, 2023. www.gendergp.com.

CHAPTER FIVE: CREATING WEBSITES

9. Courtney Quinn, "Color Me Courtney's Courtney Quinn Shares How She Built Her Personal Brand," *Teen Vogue*, September 24, 2018. www.teenvogue.com.

FOR FURTHER RESEARCH

BOOKS

Raina Burditt, *Scratch Programming for Beginners*. New York: Rockridge Press, 2020.

B. Keith Davidson, *Web Developer*. New York: Crabtree Publishing, 2022.

Clara MacCarald, *Gig Jobs in Gaming*. San Diego, CA: BrightPoint Press, 2023.

INTERNET SOURCES

"Learning for Ages 11 and Up," *Code.org,* n.d. https://code.org/student/middle-high.

"Electronic Games," *Britannica Kids*, n.d. https://kids.britannica.com.

"Free Coding Projects," *National Geographic Kids*, n.d. www.natgeokids.com/au/kids-club/cool-kids/general-kids-club/free-coding-projects/.

WEBSITES

Hour of Code

www.hourofcode.com/us

Hour of Code is a part of the organization Code.org. Its website offers coding activities and lessons for students of all ages.

Scratch MIT

www.scratch.mit.edu

Scratch MIT is a coding website run by the Massachusetts Institute of Technology. It allows people to practice basic programming. Students use blocks of text and images to make programs.

Wix

www.wix.com

Wix is an online website building platform. Users can create their own websites for free. The site also has a blog where people can find tutorials, web design tips, and other useful information.

INDEX

IMAGE CREDITS

Cover: © Monkey Business Images/Shutterstock Images
5: © SeventyFour/Shutterstock Images
7: © FilmStudio/iStockphoto
9: © BAZA Production/Shutterstock Images
11: © Sladic/iStockphoto
12: © Pentao10/Shutterstock Images
16: © AlesiaKan/Shutterstock Images
18: © Drazen Zigic/iStockphoto
21: © Photo Beto/iStockphoto
22: © Wachiwit/iStockphoto
24: © SeventyFour/Shutterstock Images
27: © MesquitaFMS/iStockphoto
28: © ot.sun/Shutterstock Images
31: © DragonImages/iStockphoto
33: © BSD Studio/Shutterstock Images
34: © senrakawa/Shutterstock Images
37: © Narin Nonthamand/Shutterstock Images
38: © sarinyapinngam/iStockphoto
41: © zeljkosantrac/iStockphoto
43: © Casimiro PT/Shutterstock Images
44: © JJ Farq/Shutterstock Images
46: © Gorodenkoff/Shutterstock Images
49: © Jose Calsina/iStockphoto
50: © Casimiro PT/Shutterstock Images
53: © Waniza/Shutterstock Images
55: © The Attico Studio/Shutterstock Images
56: © kali9/iStockphoto

ABOUT THE AUTHOR

Christopher Forest is a middle school teacher in Massachusetts. He enjoys writing books for all ages. He has written articles, nonfiction and fiction stories, and novels. In his spare time, he enjoys watching sports, playing guitar, reading, and spending time outdoors.